Marshes

Marshes

The Disappearing Edens

William Burt

Yale University Press New Haven & London

Published with assistance from the foundation established in memory of Philip Hamilton McMillan of the Class of 1894, Yale College.

Designed by Sonia L. Shannon.
Set in Bulmer type by Tseng Information Systems, Inc.
Printed in Italy by Eurographica SPA.

Library of Congress Cataloging-in-Publication Data
Burt, William, 1948–
Marshes : the disappearing Edens / William Burt.
p. cm.
Includes bibliograhical references and index.
ISBN: 978-0-300-12229-9 (clothbound : alk. paper)
1. Marshes. 2. Marshes — North America. I. Title.
QH87.3.B87 2007
578.768097 — dc22   2006026961

A catalogue record for this book is available from the British Library.
The paper in this book meets the guidelines for permanence and durability of the Committee on Production Guidelines for Book Longevity of the Council on Library Resources.

10 9 8 7 6 5 4 3 2 1

Frontispiece: Least Bittern (pl. 8)

Cover: Spartina Tufts (pl. 1)

to Carol, in loving memory

We need the tonic of wildness, to wade sometimes in

marshes where the bittern and the meadow-hen lurk,

and the booming of the snipe . . .

Henry David Thoreau, *Walden*

Contents

Color Plates

Preface

I think of marshes as withholding, mysterious, exclusive, full of unseen goings-on.
Who knows what lurks out there among the stems and shadows — what birds build
nests, what flowers grow, what hidden beauty lies? I became enthralled first as a
teenager, by peek-a-boo rails and bitterns on a cattail island in Connecticut; then
later by other birds, in other marshes — and gradually, inevitably, by the beauty
of the marsh itself. The thrall endured. For more than thirty years now I've been
prowling marshes of all kinds, all over North America, day and night alike, with all
kinds of ungainly camera gear; and what you see here, in these photographs and
stories, is what I have to show and tell.

 Most of the photographs portray either the marsh scene — near or far — or
the mystery birds within. They are selected solely for pictorial qualities, and as
such they make no pretense to a balanced, let alone comprehensive, sampling of
marsh subjects. This volume is an evocation and an exploration, not a catalog of
marshland life, and so some plants and birds are generously represented whereas

others—indeed some entire phyla—are absent altogether, depending on my predilections and, not least of all, my photographic fortunes.

Most of the scenes were photographed with large-format view cameras, which, while cumbersome, are always my first choice, because they enable such high reproduction quality. Most of the birds, on the other hand, were photographed with specially designed 35mm equipment, as described in the concluding section, "About Photography." Some of these marsh birds, in particular the smaller rails, are among the most intriguing, most vexing, and most eagerly pursued of all bird species: nocturnal, little known, exceedingly elusive and almost impossible to see—and harder still, of course, to photograph.

I hope that this portrayal of the marsh achieves two things. First, through the travel observations and the stories, I hope to provide a true and telling look at what the state of the marshlands is today, in twenty-first-century North America. To that end I tell just what I see, and where, however grim or glorious, and I tell what others saw, the early naturalists, in days gone by.

Most of all, though, after all the searching and striving with a camera, I hope to bring the marsh to others, so they too can see some of the treasures out there—still out there—in these withholding Edens.

Introduction

How could a person *not* be intrigued by marshes? I'll never know. I've always been intrigued by them, first as a young boy lured by their hidden treasures, then later as a bigger boy — well, lured by other hidden treasures.

It's true, the marshes are invaluable resources, as the biologists remind us: they filter and clean water, contain floods, provide a nursery for fish and shellfish, waterfowl, and countless other kinds of wildlife, including nearly one-third of all threatened and endangered plants and animals. So yes, the marshes are invaluable: to us, never mind the creatures that actually live in them.

But for me, the appeal has never had much to do with such utilitarian adult concerns. I've always been drawn to marshes because they are such mysterious concealing places with this lure of the forbidden and the out-of-bounds, like the prohibited frontier beyond a little boy's backyard. And I've been drawn to them because they are so full of birds: they were the hidden treasures that first lured me, and after many years still do. No other acreage I know so artfully conceals so

facts on their surpassing size, or plant diversity, or number of nesting birds per acre, say, or ducks produced per annum. But they are certainly among the biggest and the most productive, and they are, importantly, among the wildest. These are not the managed marshes of the refuges, but the neglected ones: the uncut gems, and the relics of real marsh wilderness, so far as we still have them. They are the inspirations for this book.

PLATE 1. *Spartina* Tufts
Old Lyme, Connecticut, August 1981

PLATE 2. Slender Blue Flags
Lyme, Connecticut, June 1981

PLATE 3. Coast Milkweed and *Sagittaria*
Iberia Parish, Louisiana, May 2001

PLATE 4. Calopogons
East Haddam, Connecticut, June 1983

PLATE 5. Large Blue Flags
Near Ellsworth, Maine, June 1982

PLATE 6. Clapper Rail on Nest
Old Lyme, Connecticut, June 1974

PLATE 7. King Rail
Old Lyme, Connecticut, June 1975

PLATE 8. Least Bittern
Old Lyme, Connecticut, August 1994

PLATE 9. King Rail, in Motion
Old Lyme, Connecticut, June 1975

PLATE 10. Least Bittern, Peering
Polk County, Iowa, July 1993

PLATE 11. Least Bittern, Peeking

Squaw Creek National Wildlife Refuge, Missouri, June 1993

PLATE 12. Least Bitterns
Squaw Creek National Wildlife Refuge, Missouri, June 1993

ONE

Connecticut *Home Marsh*

Uncharted wilderness, a vestige of the Wild Frontier: that's what Grandfather's kingdom-by-the-river was to a young boy of the tended Boston suburbs. There was no end of wildness to explore: streams, green tangles, hemlock groves and open beech woods, and sheer granite cliffs with promontories sloping down into the river. Along the banks you could find arrowheads and sunning black snakes, ospreys rowing overhead, and sometimes even big bald eagles. But it was another, less accessible frontier that really stirred my teenage curiosity, out in the channel, just a few hundred feet beyond the dock yet somehow distant as the Congo: Goose Island, a neglected nowhere land of mud and tide and waving blades of cattail.

One July evening I rowed out across the channel, stepped off into the greasy mud along the bank and tied up the boat, and set out among the cattails to see what I could see. I made way to a little opening, where the mud was fissured with clear rivulets; and here, as the dusk closed in I stood, enthralled. Strange slender forms stepped out from the shadows and probed the mud like chickens, without the least regard to my unlikely presence. They were rails, the first I'd ever seen: Virginia rails and sora rails. Marsh wrens skulked among the reeds, tails cocked, and scolded with a harsh *chack-chack*. A lone silhouette came gliding by, low, its neck scrunched up like a heron's, but this bird was small, hardly bigger than a meadowlark—my first least bittern.

■ Strange slender forms stepped out from the shadows and probed the mud like chickens, without the least regard to my unlikely presence. They were rails, the first I'd ever seen: Virginia rails and sora rails. ■

I was enchanted by this shady world, and other rowboat explorations followed, up the cove and down and up the winding creeks, through narrowing corridors and ever taller reeds until all orientation in the world around was lost. I'd found my own little Everglades, and it was my calling that summer to explore them and find their hidden birds.

■

Long ago, in the last years of the nineteenth century, the marshes of the lower Connecticut were widely known as a shooting ground for the very birds I'd seen in the July dusk: the rails. In those days rails still plied the marshes in great numbers, especially in fall, and especially in the wild rice marsh upriver, at Hadlyme; and in great numbers they were shot. Skiffs were poled over the marsh at flood tide, flushing the bewildered birds and affording easy targets, thanks to the feeble, spluttering, leg-dangling flight of these most cover-dependent of all birds. Often a hundred or more would fall, mostly soras, to a single gunner on a single tide. For some, this sportsman's history remains the most notable Americana associated with Connecticut River marshes.

For me, quite another piece of history evoked the Connecticut past: the discovery in June, 1884, of the nest of a little-known bird called the black rail, downriver on a sprawling flat of salt meadow called Great Island. The man was Judge John N. Clark, of nearby Saybrook, and the nest was the third ever found. To this day, Great Island remains the northernmost nesting site ever known for the black rail on the East Coast.

Did Clark's black rails still haunt Great Island? Stirred by the thought, I set out to search his olden meadow, and while I never found his legendary rails, I did find a whole new waiting world that begged investigation: the world of the salt marsh, where fine salt-meadow cordgrass, *Spartina patens,* lay in lyric tide-tossed mats and tufts. Pretty, pampering material, that fine fluffy cordgrass, a real pleasure to stray through; and aromatic, pervading the June air with its inimitable salty-sweet bouquet. But there were other grasses, and other fetching salt-marsh plants: black grass, *Juncus gerardi,* heavier and darker than the *patens,* splotching the meadow like cloud shadows; and dainty blue-green spikegrass, *Distichlis spicata,* where the meadow is lower and wetter, more accessible to tides; and, where the marsh is wetter still, staying the front line before the tides, the coarser-bladed salt-marsh cordgrass, *Spartina alterniflora,* fringing the shorelines and even springing from raw mud. And the tall, scepterlike big cordgrass, *Spartina cynosuroides,* along creeks and channels; and various sedges (*Scirpus*), and rushes (*Carex*)—all were in vast supply.

And there were flowers, late in summer: rosemallows, gerardias, goldenrods, fleabanes, salt-marsh asters of two kinds; and the ethereal sea lavenders, tingeing the meadow with a blue-grey mist. . . . I got to know them all, and I got to know the special salt-marsh birds, the king and clapper rails, seaside and sharp-tailed sparrows. And best of all I got to know the feel of the salt meadow, not just the catalog of names but the sensory detail, and the assuring luxury conveyed by all that life and open space and grass. I've since wandered in bigger, wilder, and far richer marshes, but none again were ever quite like these, at the mouth of the Connecticut, where it was all new.

But so much for boyhood reminiscences. What of these river marshes now? How have they fared in the ensuing thirty years? Great Island, stronghold of that fluffy cordgrass?

I take a drive down Smith's Neck Road, past old stone walls and pasture grown to cedar, not yet subdivided, and I stop at the State Landing for a look. Something new, right at the water's edge: an observation deck, with weatherproofed interpretive displays. Plaques, with notes on the exhibits: *Fishes of the Tidal Marsh, Ducks—Family Anatidae* (*Over thirty species of ducks, geese, and swans breed, migrate, or winter in Connecticut . . .*), and *Terns—Sterna species. . . .* Okay, okay, some information. I look out across the channel, and the sprawling island: all marshy meadow, still. Some added nesting boxes, new osprey nesting platforms, but all pretty much the same. I look upriver, and now—at once—something does look wrong. The meadow ends, abruptly, at a solid wall—a wall of reeds, tall and imposing, like the phalanx of a marching army. Still distant, subtle to the naked eye; but wrong.

I drive a mile upriver for a closer look, and here the army is established, thick and fast, across the entire north end of the island, and all river marsh above it. Only some thin strips, along the inside shore, remain unoccupied.

Another mile upriver, now I drive down Ferry Road, to the new state Department of Environmental Protection facilities: a veritable villa, with Marine Headquarters building, laboratory and garages, State Pier, and the bristling research vessel *John Dempsey* at its berth.

The dread invasive common reed, *Phragmites australis*, overtaking native cattail marsh on the Connecticut River estuary.

Impressive. Some facilities for the public, too: a pruned park, with picnic tables and pagoda; and a classy modern boardwalk, stretching south along the river to a high-perched observation deck. I follow the boards, taking in the ease, the solid reassuring spring, the marina-quality opulence; and I climb up to the deck . . . and here it is, the confluence of rivers—the Connecticut, the Lieutenant, the Duck— and all their arrayed marshes, sprawling southward to Great Island. Marshes, meeting place of land and sea: last bastion of frontier. Timeless, aboriginal, en- during. But are they? What do I really see? I see reeds, the giants of that marching wall I saw down at Great Island, except that here they are not a wall, but an entire landscape. Here, there is no other marsh.

No, what I see here is not wilderness, nor marsh frontier, but marsh usurped: a place where real marsh used to be.

■

Phragmites australis, or just *Phragmites,* also known as common reed, is a ruthless rampant bully of a plant, a perennial that grows up to sixteen feet tall, in dense, bamboolike monoculture stands. In Connecticut, researchers have found it in peat samples dating back three thousand years, so the species is, confoundingly, a native plant, and a once rightful and respectful member of the wetland plant community.

But in the past forty or fifty years, something has gone very wrong with *Phrag- mites* on the eastern seaboard. Since about 1970 it has behaved like an aggressive alien, spreading and overtaking wetlands like unchecked disease, snuffing out the rightful plants and ousting native birds, acre after acre. At the mouth of the Con- necticut, scientists say, it has been spreading at the ghastly rate of more than 1 per- cent per year in the more saline, *Spartina*-dominated lower marshes, and more than 2 percent in fresher marsh upriver. It spreads both by seed dispersion and by extension of its rhizomes (underground stems) and stolons (aboveground stems)— and, like some creation from a 1950s horror movie, by regeneration from small cut- tings, so that mowing and chopping the plant to bits may only speed the spreading. *Phragmites* moves inexorably, displacing all in its way, a true floral juggernaut.

cattails, bulrushes, fleabanes, rosemallows; birds reappearing. And the project will continue, Capotosto says, here and elsewhere in Connecticut. Within the decade, he feels, his team should have the entire estuary largely under control. He is optimistic. And committed.

"We're in it for the long haul," he says.

That is good news for the Connecticut River estuary, and it comes none too soon. How pitiable these Old Lyme marshes, now. They seem little more than pretty frills, fringing their river towns like furbelows, or doilies, and I say this not because I view them with more traveled or more jaundiced eyes, but because they *are* diminished Edens. Three-quarters of those cattails were already gone when Capotosto's team began, and the rest would surely follow soon without this rescue operation; and it would not be long, I fear, before the last lone rail flapped off to seek home elsewhere.

What if the phrag control were discontinued? What would the picture be in yet another thirty years?

I can see pleasure boats and jet skis, well-kept lawns, tame swans and mallards begging bread. And reed, everywhere the silent reed.

Where would a boy go then?

PLATE 13. Red Maple at Marsh Edge
Lyme, Connecticut, October 1976

PLATE 14. Sweetflags
Hadlyme, Connecticut, May 1978

PLATE 15. Sweetgale and Sedges
Pittsburg, New Hampshire, September 1978

PLATE 16.   Swamp Rosemallows
Old Lyme, Connecticut, August 1981

PLATE 17. Marsh Marigolds
Colebrook, New Hampshire, May 1979

PLATE 18. Silverweed

Old Lyme, Connecticut, May 1981

PLATE 19. Perennial Salt-marsh Asters
Old Lyme, Connecticut, September 1981

PLATE 20. Virginia Rail on Nest
Old Lyme, Connecticut, June 1977

PLATE 21. King Rail on Nest
Old Lyme, Connecticut, June 1975

PLATE 22. Nest and Eggs, King Rail
Old Lyme, Connecticut, June 1975

PLATE 23. Sharp-tailed Sparrow
Old Lyme, Connecticut, June 1974

PLATE 24. Least Bittern, Peeking Out
Polk County, Iowa, July 1993

PLATE 25. Least Bittern, Wings Out
Old Lyme, Connecticut, August 1994

PLATE 26. Least Bittern and Cattail Blades
Old Lyme, Connecticut, August 1994

TWO

To Maryland *and a Big-league Meadow*

Heading southward from Vienna, Maryland, you come into flat, wide-open farming country. You pass neat plats with bounding lines of trees, abandoned houses, collapsed sheds and fields lain fallow, logged-over lots with bone-bare standing snags, and vultures perched in stark still-lifes. Bleak, unfriendly country. You pass unfriendly signs, with rusted bullet holes: *No Trespassing. Beware of Dog. No Hunting.* You recall stories: locals confront visitors, at gunpoint.

The country closes in with loblolly pines, dark and forbidding. Some marshy openings appear, and some drowned trees, bone-white and bleak; and more unfriendly signs, with bullet holes.

Then at last you break out into friendly country, to the sight of creeks and pools and sprawling green: mile after mile of sprawling salt-marsh green. You're in Dorchester County, Maryland, on the Eastern Shore of Chesapeake Bay, and you're headed out across a marshland unlike any other on the eastern seaboard, the Elliott Island marsh. This is no snug, New England kind of marsh; this is a

irresistible to many wildlife bureaucrats and civil engineers, and one that may be therapeutic in the case of marshes on the arid western plains, say, where containment helps conserve what precious little water has not already been diverted for irrigation. But the practice is meddlesome and pointless, even counterproductive, when applied to pristine tidal coastal marshes such as this one. In fact one of these structures, a three-mile ring dike, was built at Elliott some years ago, with the help of Ducks Unlimited; and whenever Armistead has walked around this dike, he says, he has found "almost nothing, literally, inside of the dike, all the birds being on the outside." He notes a similar situation at nearby Deal Island Wildlife Management Area, where another prime piece of tidal marshland has been diked. Birdlife burgeoned inside that dike for the first few years, he recalls, but since then it has rapidly and steadily declined.

Armistead has one other concern for Elliott, and it's a big one: *Phragmites,* the metastatic reed that has infested river marshes in Connecticut and elsewhere in the Northeast. It's still a minor player here at Elliott, he says; but it has established some substantial beachheads, particularly at the northern end, where it is extending fingers from the roadside ditches out into the switchgrass, the cattails, and the Olney's three-square, even out into the pure *Spartina patens* marsh itself, the very heart and essence of the marsh and home of the black rail. That scares me, more than the burning, the building of dikes, and even the effects of rising sea—more than all these put together, really—because I've seen what the *Phragmites* juggernaut can do. I've seen it spread and overtake whole marsh communities, whole landscapes, and transform a small but vital estuary system into a near-monoculture waste, not unlike the one at Hackensack, New Jersey, just south of New York City.

Soon, I fear, without continued and relentless effort by the state control teams, the last of the wild Connecticut River cattail marshes would succumb. And that would be a tragedy, of course. But it would be a minor tragedy—a minor-league tragedy—compared to what could happen here, to a big-league marsh in Maryland.

Here, at Elliott Island marsh, there is so very much more to lose.

PLATE 27.   Dawn
Elliott Island, Maryland, October 2000

PLATE 28. Three-square and Spikegrass
Elliott Island, Maryland, November 1999

PLATE 29. Nest and Eggs, Laughing Gull
Near Wachapreague, Virginia, May 1997

PLATE 30.   Salt-marsh Fleabane and *Spartina*
Elliott Island, Maryland, November 1999

PLATE 31. *Spartina patens*
Elliott Island, Maryland, August 2001

PLATE 32.   Moon and Pines, Marsh Edge
Elliott Island, Maryland, September 1993

PLATE 33. Black Rail in *Spartina*
Elliott Island, Maryland, June 1985

PLATE 36. Sea Pinks
Elliott Island, Maryland, August 1992

PLATE 37. Spikegrass and *Spartina*, Early Light
Elliott Island, Maryland, October 1999

PLATE 38. Marsh Ferns and *Spartina*
Elliott Island, Maryland, September 2001

PLATE 39. Black Rail, Emerging
Elliott Island, Maryland, July 1985

PLATE 40. Black Rail at Nest
Elliott Island, Maryland, July 1985

THREE

Manitoba *Sedges*

After zigzagging through the little town of Douglas, Manitoba, Route 340 beelines south across a marshy basin, which sprawls soft and green for a long ways to either side. To the west, this sea of sedge extends about two miles, then merges into solid grassland specked with cattle; and to the east it extends farther still, and not to any such pastoral destination but reaches ever more wild, more vague, and inaccessible. For five miles there is nothing but the waving sedges, huddled willow shrubs, and the odd slow stream or pool.

Then meadow merges with the deeper, darker green of tamarack and sphagnum, and a mile-long lake appears, Sewell Lake, clear blue and pristine, ringed by meadows traced with trails of moose and elk. Then other lakes appear, with their thin bands of marsh: blue holes in the blanket of green tamarack. Here is the sort of wilderness a person dreams about, unsullied by the centuries and seen by few human eyes.

61

never really treacherous, perhaps, but surely nasty. Dead Man's Swamp, for example, with its tall lacerating blades of bulrush, and a floating mat that gurgles as it sinks beneath you, and its reputed "springs."

But again, some marshes can be soothing, pampering, pleasant as the village green; and this Douglas, Manitoba, marsh is one of them. No place is more salutary to the senses—except, of course, on those dark nights when you're out there wading in the sedges, waist-deep and alone, and the uncertainty of quicksand comes to mind. . . .

But then what would a marsh be, what would any wild place be, without uncertainty?

PLATE 41. Nest and Eggs, Black Tern
Douglas, Manitoba (Canada), June 1994

PLATE 42. Sensitive Ferns
Near Pembroke, Maine, July 1995

PLATE 43. Watershields

Moosehorn National Wildlife Refuge, Maine, July 1995

PLATE 46.   Yellow Rail, Peeking
Douglas, Manitoba (Canada), June 1987

PLATE 47. LeConte's Sparrow
Douglas, Manitoba (Canada), June 1994

PLATE 48.   Yellow Rail
Benson County, North Dakota, June 1991

PLATE 49. Sora
Douglas, Manitoba (Canada), June 1987

PLATE 50.   Fireflies and Lightning
Douglas, Manitoba (Canada), June 1993

PLATE 51. Sedge Wren Singing
Benson County, North Dakota, June 1992

PLATE 52. Fireflies over Marsh
Benson County, North Dakota, June 1992

FOUR

South *and Along Coasts*

"The Atlantic salt marshes," wrote the southern naturalist Brooke Meanley, "are the last frontier of the Eastern United States."

It's true, and for the marshes on the Gulf as well. All other coastal lands have long been used and reused, shaped and reshaped by the hand of man, but not the coastal marshes, by and large; they are much the same today as they were when they first emerged, centuries and tens of centuries ago. And they are vast. The most extensive marshes anywhere in North America, by far, are the coastal marshes of the Gulf and south Atlantic states.

Yet I'd seen little of that "last frontier." So I planned a tour, a southern expedition. It was high time. I'd begin in Texas, head east along the Gulf to Florida, and continue north up the Atlantic seaboard to New Jersey, ultimately, just short of the megalopolitan Northeast.

I stopped first at the noted Texas refuges—Annahuac, Brazoria, San Bar-

nests, and for the morning I was one of those old naturalist-explorers of the past, an A. C. Bent out in the teeming sloughs of North Dakota, say, or a William L. Finley on the early marshes of the Klamath. Yes, I know, these were only laughing gulls, undiscriminating riffraff equally at home in coastal dumps or shopping malls, or fast-food parking lots; but out here on the wild free spaces of the marsh, where they go to nest, their numbers were exhilarating. And you'd never even guess that these black-hooded breeding-plumaged beauties, at home on their breeding marsh, could be the same garbage-picking sleazes that you see off-island.

And let me say, my island-of-the-past illusion was not all illusion. These ocean marshes of Virginia come as close as any on the continent to what the naturalists of old once saw.

Other birds, not just the schizophrenic gulls, live in these vast salt marshes. Clapper rails lurk in the tall *alterniflora* along tidal guts, and canopy their basket-fuls of eggs against the pirate eyes of fish crows, and no doubt the gulls, and other cruising villains overhead. And with the dingy rails live dingy seaside sparrows, stuffing their small nests under the dead-grass wrack washed up by tides. Bolder birds, such as Forster's terns and willets — sometimes even beach-loving oyster-catchers and black skimmers — place nests directly on the wrack, in plain view.

I saw many shorebirds that May day: about a hundred whimbrels traveling high together, calling, and groups of yellowlegs and dunlins, black-bellied plovers, knots, turnstones, and many of the smaller "peeps," veering past in silver-flashing schools and settling on the flats and vanishing, like smoke. But all these were only transients, intent on other wetlands far away.

Except for Elliott Island marsh, nearby in Maryland, this salt-marsh com-plex in Virginia is, to me, the finest wetland wilderness in eastern North America. But I found another salt-marsh beauty farther north, up in New Jersey. It is much smaller, and more proximal to man, of course — and it holds nothing like the teem-ing life of the Virginia marshes — but it is within its limits just as much a wilder-ness, and just as much an Eden.

New Jersey, did I say?

Home of Atlantic City and casinos, toxic dumpings, gangland murders? And Elizabeth refineries, those sci-fi landscapes of steel pipes and tanks and stacks dispensing smokes into an already amber sky? And nearby Hackensack, once home to a real marsh, Hackensack Meadows, and now home to a stadium called the Meadowlands? And to pollution, landfill, and *Phragmites?*

No, I hadn't had the best impression of New Jersey. But then I discovered the marshes at Great Bay, near Tuckerton.

■ *Five miles* of open salt marsh—in New Jersey, just a two-hour drive from New York City and the Bronx. Five miles of open space, and grass, and looping lines of creeks, and pools. . . . ■

I found no *Phragmites,* no toxic dumps, nor anything else unseemly there that late October afternoon, only mud and cordgrass, creeks and pools, and a horizon-to-horizon sweep of golden grass as pretty as I'd ever seen. A little road runs out across, spanning creeks on one-lane bridges, passing two or three marinas, and continuing across unbroken marsh for several miles, dead-ending finally at Little Egg Inlet, on Great Bay. In all, the road traverses some five miles of open salt marsh.

Five miles of open salt marsh—in New Jersey, just a two-hour drive from New York City and the Bronx. Five miles of open space, and grass, and looping lines of creeks, and pools . . . intriguing pools. Unusual. They cluster in close groups, curving, bulging and extruding into inkblot squiggles, and connecting playfully in Byzantine arrangements. Where else, this fancy show of pools and mazes? This doodling interplay of land and sea? I'd never seen a salt marsh like it.

■

I came back one damp November day to find the whole marsh shrouded in a fog, transmuted, a place of perfect stillness penetrated only by the outside call of some woebegone yellowlegs, or gull, or wayfaring flock of dunlins. In fog, more even than the dark of night, I think, a marsh is an enchanted place; and safe, insulated for the while against the noise and nonsense and unsightliness of worlds beyond. The insulation is but vapor-thin, of course, and transitory, but for a person there within, in his own safe private time and place, it is reality enough. I reveled in it, that November day.

■ In fog, more even than the dark of night, I think, a marsh is an enchanted place; and safe, insulated for the while against the noise and nonsense and unsightliness of worlds beyond. The insulation is but vapor-thin, of course, and transitory; but for a person there within, in his own safe private time and place, it is reality enough. ■

PLATE 53. Lily Pads and Bladderworts
St. Marks National Wildlife Refuge, Florida, May 1997

PLATE 54. Prairie Marsh
Cameron Prairie National Wildlife Refuge, Louisiana, May 1998

PLATE 55.   Live Oaks, Marsh Edge
St. Simon Island, Georgia, May 2001

PLATE 56. Millet and Pool
Brigantine National Wildlife Refuge, New Jersey, October 1999

PLATE 57. Evening Marsh and Pools
Deal Island, Maryland, October 2000

PLATE 58. Salt Marsh Detail
Great Bay, near Tuckerton, New Jersey, November 1999

PLATE 59. Low Tide
Coosaw River, South Carolina, May 2000

PLATE 60. Salt Marsh and Pools
Great Bay, near Tuckerton, New Jersey, May 2000

PLATE 61. Salt Marsh in Fog
Great Bay, near Tuckerton, New Jersey, November 2000

PLATE 62. Roseate Spoonbill
Everglades National Park, Florida, March 1986

PLATE 63. Purple Gallinule
Everglades National Park, Florida, March 1986

PLATE 64. Laughing Gull

Near Wachapreague, Virginia, May 1997

PLATE 65. Common Moorhen
St. Marks National Wildlife Refuge, Florida, May 1997

PLATE 66. Alligator, Loafing
St. Marks National Wildlife Refuge, Florida, May 1997

FIVE

West *and Water*

One

One cozy Christmas morning when I was a boy I received a hulking book, called *Birds of North America*. It was a reissued work, originally published in 1917, and it was something of a hodgepodge, with a mix of classic color paintings by Louis Agassiz Fuertes, old photographs, and text accounts by long-departed naturalists such as T. Gilbert Pearson, Edward Howe Forbush, and the old nature sage himself, John Burroughs.

The writings were not only quaint, but hopelessly outdated, as even a fourteen-year-old could see; and the old photographs of birds, all antique black-and-whites . . . well, they were evocative, at best, when not downright amusing. Some were photographs of subjects plainly shot and stuffed and wired in place, like prepared museum exhibits; some were of young birds recruited from the nest, and lined up

their surrounding tules stiff and mummified. Not a good year, 2001, to see Finley's Malheur Lake. I'd picked the region's worst drought year in two decades, and one of the worst in half a century.

■

Not a good year to see the Klamath Basin either, I suppose. But here I was, in Oregon, committed, so I drove on down to Klamath Falls; and yes, there would be water troubles here, all right. I learned it in a hurry when I stopped for gas, and asked the attendant about motels ahead at Tule Lake, down near the refuges.

"Oh, there *might* be something still down there," she said dismissively. "But maybe not. They're all moving out, because the farmers can't get any water. It's all going to save some *fish.* . . ."

She had war in her eyes, and I dared not ask her how the fish were doing. Or the marshes, or the birds.

The trouble is, as it was neatly nutshelled in the *New York Times,* there are just "too many claimants for too little water." The claimants are (1) the farmers—whose farmland, once full of lakes and marshes, arguably ought not to have been drained and farmed in the first place; (2) the fish—which include two species of federally endangered suckerfish and, downstream, the threatened coho salmon; (3) the area's native American tribes and commercial fishermen, downstream—all, dependent on the salmon; (4) the refuge birds—including 80 percent of all the waterfowl using the Pacific Flyway, and the largest wintering bald eagle population anywhere in the United States except Alaska.

Legally, the first priority goes to the endangered species, which is to say the suckerfish; the second goes to the native American tribes; the third to agriculture; and the fourth and last to nonendangered wildlife, and the refuges. Effectively, though, because the refuges are last in line for water, so too are populations of the endangered fish—which inhabit two of the refuges, Tule Lake and Clear Lake.

Showdown was inevitable. The scene was set for trouble long ago, with the Klamath Reclamation Project's promise of cheap land and water, and the ensuing rush to settlement. All it took to spark a crisis, after years of free and easy irrigation and depletion, was the addition of two factors: the recent legal recognition of new "users"—the native Americans and the native endangered fish—and, finally, a year of major drought. The drought of 2001, let's say: the year of my visit. According to U.S. Fish and Wildlife, water had become so low in Upper Klamath Lake, the basin's major reservoir, that any further decline would directly jeopardize the two endangered suckerfish (and the coho salmon runs, downstream); so the Bureau of Reclamation felt it had no choice and closed the headgates, leaving ruined crops and angry farmers. So angry were some farmers, initially, that at least one covert effort was made to reopen the headgates.

Into this embattled region, then, where food and livelihoods and lives of species were at stake, I drove south through hills and furrowed farmlands to the northern California town of Tule Lake, then westward into refuge country, through a gauntlet of hand-scrawled signs along the road: *Call 911. Some Sucker Stole Our Water.*

Sarcastic, bitter, angry signs. Cries for attention, commiseration. You could almost hear the voices, see the faces: *Federally Created Disaster Area.* And *Water for Farmers, Not for Suckers.*

Not a comfortable situation. I was an outsider here, a snooper in a troubled land. Homes and happiness lay in the balance, and here comes along this lone outsider, touring fancy-free and searching for his pretty marshes and his pretty pictures, and his traces of the gauzy past.

But it was my pretty marshes that belonged here, was it not? The water, too, and all the birds and fishes, even silly-sounding suckerfishes, they belonged here too; and they had been here for some time, had they not? And a good deal longer than the farmers? My sympathies reversed, when I recovered from the onslaught of those signs. Other lives hung in the balance too, after all; and the lives of actual species, entire kinds. They need water too and they, unlike the farmers, cannot pick up and move and live another kind of life. Either they live here or they don't live. And this is not a country meant for sugar beets and onions and potatoes, moreover; it's meant for tules, and fish, and pelicans and eagles. And ducks, geese, grebes, and about 220 other bird species listed for the Klamath Basin, not to mention countless other kinds of living things. . . .

Further provocation was forthcoming, as if it were needed, from a shady lawn beside the road, not two hundred yards from the refuge headquarters and visitor center. Here, surely, was the sign to end all homemade signs: a sarcastic master-piece that must have cost someone a lot of trouble, time, and money. It was a big, all-metal job supported by huge posts and painted with a counterfeiter's care, white-lettered on official outdoor-recreation brown, in painstaking simulation of the signage posted on all federal parks and refuges and other public lands: *Please Thank the U.S. Fish and Wildlife for Destroying the Ecosystem and the Economy of the Klamath Basin.*

So, the U.S. Fish and Wildlife Service. It's their fault? I was all too mindful, reading this, that it was for farmers that these marshes had been bled and butch-ered in the first place, so the water could be piped away to irrigate and plant the desert—and the temptation, here, was to be sarcastic in return. Take *that*, sarcastic farmers!

But my mission was to see the lakes and marshes that remained: the 25 percent. What would Finley and Bohlman find today? At Tule Lake, my first destination?

They wouldn't recognize the place, wouldn't even find those waterways and tule mazes of a century ago. They would find a lake, alright, but just a lake; just water lapping at a frothy shore, and dikes, and flooded pasture with emerging weeds. No tules here, not any more. Not at Tule Lake.

At Lower Klamath Lake? Finley's onetime "land of dreams," where he and Bohlman cruised by boat and camped on matted tule islands? They wouldn't rec-ognize it either, I'm afraid; they would find only a manufactured complex of neat dikes and ponds, square ponds, laid side by side like sewer pits, or fish farm ponds.

I drove along the dikes, hoping for some traces of the tule islands, but the only tules here were fencelike rows along the dikes, in ditches, and then some isolated dusty stands, long since marooned and dead of thirst: only the desiccated husks, still bowed in place. Some of the ponds were dead dry too, their beds baked hard and crazed, and crusted white with alkalai. In one pond two monster machines were groaning away, and moving earth around.

Only two of these diked units at the Lower Klamath—along the highway, for all public eyes to see—held any real water, and waterfowl. They held real ducks, hundreds of them, maybe a thousand ducks or more; but these ponds were missing something. And the ducks were missing something: they were ducks without a flora, without any gracing tules, sedges, cattails, or floating-leaved aquatics—without any living ambience or cover whatsoever, as if set free in a big swimming pool. They had their official designated "wetland," I suppose; but they did not have marsh.

But does anybody care? Does it matter that the real marsh here has been replaced by water-regulated compounds that resemble borrow pits, or excavated fish farms? Not here, apparently. What matters here is ducks, and duck production.

"The lure of the marsh was in its *wildness*"?

Sorry, Mr. Finley, but what seems to matter now is numbers.

■

I found one hopeful outpost in the Klamath Basin, about sixty miles north of Lower Klamath and Tule Lakes. It's called Klamath Marsh, and it took me by surprise one frosty autumn morning. I was driving through this high, dry country of dark pines, and here it was, suddenly, this wide-open swathe of tules, sedges, and mirror-perfect creeks and pools, about ten miles across and wild and undisturbed as it could be, except for the lone road that shoots across. No digging or diking here; no segregated ponds, no water-pumping stations; nothing but the frosted marsh and distant streaks of pine, and fir, and the blue sky, and distant mountains. And the perfect pools, jarred only by the bobbing up of pied-billed grebes. Other birds were there, no doubt, unseen and silent in the marsh; but what would they be? What would a person hear, on a soft spring evening? Marsh wrens, soras, yellow rails?

Here it is then, Mr. William L. Finley, a bastion of your Klamath Basin marsh that "defies civilization." Still.

Two

On this high note I moved on, westward from the Klamath Basin and across the Cascade Range, then southward on Route 5, straight down the middle of the Sacramento Valley. To either side lay blue-grey banks of mountain ranges, dark and ominous, like the cloud banks of advancing storms: the Coast Range to the west and the Sierras to the east. Between them stretched the valley floor, fifty miles across and flat and cultivated as could be: every last tillable acre, mile after southward mile, from valley wall to valley wall.

■ And the remaining wildlife? About three million waterfowl still use the Pacific Flyway, and they have to spend the winter somewhere. Where do they go? ■

Once, these flat farmed lands of the Central Valley—the Sacramento and San Joaquin combined—were strewn with lakes and marshes, four million acres of them, covering almost one-third of the entire valley. These wetlands were both permanent and seasonal, and supported birds and fish and other wildlife in stupendous numbers, including wintering populations of more than half of the thirty-five million birds then using the Pacific Flyway. But no more. Settlement and cultivation has absorbed them all—and most of the valley's river water, too, through the Central Valley Project of the 1930s and 1940s, which dammed it, canalized it, and dispersed it over thousands of square miles for irrigation use.

And the remaining wildlife? About three million waterfowl still use the Pacific Flyway, and they have to spend the winter somewhere. Where do they go?

They go to the only wetlands allocated to them, the national wildlife refuges. A string of these reservations is maintained in the Central Valley, expressly for these waterfowl dispossessed.

I pulled off Route 5 and into Sacramento Refuge, the valley's northernmost and the first to welcome the arriving southbound autumn skeins, but this southbound visitor found a chilly welcome: a vending machine, at a modern auto-pay station. I did as directed, put in my three dollars and took out my official dated ticket, high-tech California style. Wildlife refuge, or amusement park? Oh well, I was invested now. Might as well go take my ride on the wildlife drive.

Signs, almost at once: *Speed Limit 20.* And *Visitors Must Remain in Their Vehicles Except at Designated Park-and-Stretch Locations.*

Park-and-Stretch Locations. Well, they do make sense here in this Central Valley, where alas there is no water room for birds except upon these ponds. But the reality was a hard one. No water room for birds? Here, over these *thousands of square miles* of cultivated valley, *one-third* of them once under lakes and marshes— no room left except on these provided ponds? Imagine.

And here they were, the ponds, square ponds; and elevated ponds, no less, set not in but on the surface, so you can sit tight in your car and watch the swimming, diving, and dabbling of the ducks at your eye level, like fish in an aquarium.

More signs: *Only Vehicles Beyond This Point.* And *Wildlife Only Outside of Vehicles.* And again, *They'll Be Out Here If You're In There.*

Okay, I understand; these reservation ponds are all that these birds have, and they just cannot be disturbed. But it's hard to see these remnants of wild marsh reduced to live exhibits-at-a-distance, to be viewed from your encapsulating car. Not a thing about the place was wild, except the waterfowl itself; even the live vegetation was inapt and unconvincing—merely upland weeds and grasses, mostly, and clumps of cattail set in neatly, periodically, like ornamental shrubs on office lawns. I suspect this project might have even been a showpiece, and a proud example of the modern re-created marsh.

"Re-created marsh": another of those euphemistic insults to the language and bereft of native marsh and unprepossessing as a military base in disrepair, but it has its working infrastructure, and its birds.

◼

The last of my mountain marshes, in southern Colorado, was Monte Vista Refuge, a neat, well-tended grassy complex where the ponds and marshes are contained and well-defined, like water hazards on a golf course. The Montes here were smaller and more distant than those looming titans at Bear River, and the birds were far, far fewer. A posted sign, *Spring Creek,* had something to say: "Spring Creek once flowed about 8,000 gallons of water through this area, creating a wetland haven. . . . Many wells were drilled to pump water for farms, ranches, and the refuge. In the mid-1960s, the spring ceased flowing. Water for wildlife is now maintained by pumped wells and irrigation ditches."

The old story, once again. You see it all across the desert West: the water is diverted, spent, and yet another wetland paradise goes dry. Ruin, or near ruin, is followed by regret, then a great expensive flailing effort to undo the damage and restore some vestige of what was. At best, some vestige.

Is there not some surviving Eden somewhere out there in the West—just a partial Eden, maybe? Just a pristine piece?

PLATE 67. Creeks and Salt Marsh, from Air
Wachapreague, Virginia, September 2000

PLATE 68. Algal Mat

Great Bay, near Tuckerton, New Jersey, October 1999

PLATE 69. Sod Bank, Marsh Edge
Near Wachapreague, Virginia, May 1997

PLATE 70. Evening, Tidal Creek
Near Onley, Virginia, November 2001

PLATE 71. Sweetbay and High-tide Bush, Marsh Edge
Saxis, Virginia, September 1993

PLATE 72. Rushes
Bear Lake, Utah, June 2000

PLATE 73. Evening Sky
Bear River Refuge, Utah, July 2000

PLATE 74. Sedge Meadow and Stream
Yellowstone National Park, Wyoming, October 2001

PLATE 75. Snow and Sedges
Yellowstone National Park, Wyoming, October 2001

PLATE 76. Yellow Pond Lilies
Elk Mountains, Colorado, July 2002

SEVEN

Saskatchewan *Plains, Sloughs, and a Certain Eden*

At last I was on my way, high in the cab of a jouncing pickup, eyes on the horizon. I'd waited a long time to see what lay out there along the prairie trail, beyond the rain pools; it had been a year since I'd made the drive in my ungainly Oldsmobile, only to be stopped short by those pools. Now, in my rented 4 × 4, there would be no stopping me.

I bumped and splashed on through the pools, chin forward in anticipation. Every eyeful now was new, every stretch of plain and rolling hill and marshy hollow, each flitting mix of sparrows, pipits, longspurs . . . but the real lure lay yet ahead, in the thin distance: the Eden that I'd read about. Was it still there?

PLATE 78. Bulrushes, Evening
Near Saskatoon, Saskatchewan (Canada), June 2000

PLATE 79. Spikerushes
Near Eyebrow, Saskatchewan (Canada), June 2000

PLATE 80. Reeds and Water, Sundown
Crane Lake, Saskatchewan (Canada), July 1997

PLATE 81. Sedges and Pools
Near Brandon, Manitoba (Canada), June 2000

PLATE 82. Pied-billed Grebe
Crane Lake, Saskatchewan (Canada), June 2000

PLATE 83. Young Franklin's Gull
Crane Lake, Saskatchewan (Canada), June 2000

PLATE 84. Pied-billed Grebes
Crane Lake, Saskatchewan (Canada), June 2000

PLATE 85. White Pelican
Crane Lake, Saskatchewan (Canada), June 2000

PLATE 86. Western Grebes
Crane Lake, Saskatchewan (Canada), July 1995

PLATE 87. Eared Grebes

Crane Lake, Saskatchewan (Canada), July 1995

PLATE 88. Duckling Blue-winged Teal
Crane Lake, Saskatchewan (Canada), July 1995

PLATE 89.   Young Pied-billed Grebe
Crane Lake, Saskatchewan (Canada), June 2000

PLATE 90. Muskrat
Crane Lake, Saskatchewan (Canada), June 2000

Afterword

What have I learned, for all these travels and investigations? How do our marshes fare one hundred years and more since the early naturalists first ventured into them and reveled in their teeming birds?

I saw many blighted marshes, to be sure: blighted by the bully plant, *Phragmites;* by landfill; by drainage schemes for "reclamation"; and by almost every other human means imaginable, from ad hoc dumping here and there to wholesale agricultural conversion, as in most of Iowa, and California's Central Valley.

But that's only half the story. I saw disappeared and disappearing marshes but I saw some nearly pristine jewels, too, still wild and free and brimming with the birds and flowers that belong. We have a number of these jewel marshes, big and small alike, in North America. Some of the biggest are those explored here: the Maryland *Spartina* marsh of chapter 2, the Manitoba sedge meadows of chapter 3, and so on, concluding with the mythic lake of the last chapter, once a bird man's glory of the Northern Plains. No doubt we have still other sprawling glories, somewhere, that would hold a naturalist equally in thrall.

And we have many lesser glories—lesser in size, that is, but not in bounty, or in beauty: many of the potholes on the Northern Plains, for instance; and many of the fine salt marshes still intact along the Gulf and Atlantic coasts, such as the golden grass surprise of chapter 4, near Tuckerton, New Jersey.

And we have many rich and varied fragments in the least auspicious places, sometimes, but precious and alluring nonetheless. Some of these photographs were taken in the tiniest of marshes, even in small pools and ditches by the road on some occasions.

So just look at what remains, if you'd despair at what's been lost. Riffle through these photographs and note the colors and the textures, the detail, and above all the variety of the life you see; and keep in mind that it's all life and color of the present, not the past: it is still there. And it's just a sampling. These pictures represent a hefty mortal effort, you can be sure, but they give you no more than an inkling of what's out there in the marsh. They are mere glimpses.

I hope they are intriguing glimpses, though, and that together they convey the feeling, if not the fullness, of these Edens we still have.

About Photography

Like all things in the natural world, the subjects of these marshland photographs fall into two essential categories: those that can fly, swim, run, hop, or slink away, and those that can't.

For those that can't, I've used large-format (4 × 5) cameras, primarily for the inherent sharpness and exquisite detail, but also for the more contemplative and careful process that their use requires: you don't shoot a picture, you *compose* it on a five-inch groundglass under a black cloth. I've used both the monorail "view" and the flatbed "field" types, and have come to prefer the lighter and more compact field type for its obvious logistical advantages in places marshy and precarious. The movements of a field camera are more limited than the view's, but the range of its front and back tilts in particular, which I use most, are almost always more than sufficient for my work in the natural world.

I usually prefer a lens of "normal" focal length (a 180mm Schneider Symmar-S, or Apo Symmar), for its tastefully natural perspective, and absence of intruding optical effect. But sometimes, when the reach of a longer focal length is indispensable, I use a modest telephoto: a 360mm Nikkor-T. And I sometimes use a modest wide-angle lens—a 90mm Schneider Super Angulon—though only rarely, because I dislike the stylistic, even flamboyant perspective of most wide-angle lenses. The effect can call attention to itself, like a writer's fancy prose, and it often comes across to me as a sort of optical showing-off.

My preferred sheet film is plain old Ektachrome 64, though I've sometimes used the faster 100-speed films. I've tried the other 64-speed films on occasion, made comparisons, and almost always come back to the Ektachrome, primarily for its moderate contrast and its understated rendering of greens.

The only filters I use are graded neutral-density filters, and I use those only to control large areas of overwhelming brightness, such as skies.

Occasionally, I've used the 35mm format for photographing "landscapes," but only when there was no other way: when I wanted to photograph a dusk scene full of blinking fireflies, for instance, and no large-format lens would have the speed (maximum aperture) to record even a trace of those weak lights.

■

For photography of animated subjects, primarily birds, I've used ordinary 35mm systems, but by two unorthodox and very specialized means: close approach on foot, with a specially designed outfit incorporating multiple diffuse flash units; and not-so-close approach, by water, with long lenses from a floating blind.

The first approach, with diffuse flash, evolved in the course of photographing two nocturnal marsh birds, the black rail and the yellow rail. To this end, after many trials and many, many errors, I devised a system employing multiple flash units, as many as four, together with diffusion screens, camera, power winder, and a macro lens of modest focal length (50 to 200mm)—and of course a flashlight to see and focus by—all mounted to an aluminum frame, with sturdy struts and an extending shoulder brace. Admittedly unwieldy, and a little strange, not always easy to explain to passersby, the contrivance was nonetheless portable, and with it I could manage (usually) to negotiate a marsh at night. And, while portable, this outfit could be depended on for lighting as fine as any studio could produce. Fine light means *soft* light, and that's what the diffusion screens were for: nothing is more distressing photographically, I feel, than the hardened glassy look of raw, unsoftened flash, which can render living birds as lifeless as ceramics. But the system gave substantial depth of field, too—crucial, in my view, to fine environmental bird photography. The bird's surroundings matter too, after all; and far too many photographs abandon that big piece of the picture, in effect, by casting it into a mushy, obfuscating blur.

With this outfit then I went to work on the two rails, improving and refining it from time to time. The beauty of it, again—and only slowly did I learn to appreciate this fully—was that it allowed mobility *and* it assured good lighting, so I was free to stalk birds at large, away from the confines of the nest vicinity, where I could catch them doing things more interesting than sitting on eggs or tending young: singing, for instance, or half hiding, or creeping stealthily.

There was one hitch in this "portable studio" approach, however, and it was a major one. In order to light the subject properly and retain that precious softness of the flash, I needed to get close—very close—within four to eight feet—and of course my subjects were not amenable to that proximity. So I had to employ some special strategies, such as stalking in the dark at night, or setting up near nests or singing perches.

My second means for photographing birds of marshes, in particular the swimming and diving birds of prairie sloughs, was with a long lens from a floating blind. For that pursuit I built a modular, transportable assembly incorporating three pontoon sections and a four-part aluminum frame, with a "cowcatcher" out in front to deflect reeds and other flotsam, and a camouflaged fabric cover to conceal the photographer within. Other features included an adjustable camera post, flash mounts, and waterproof storage boxes. I should say that this mobile blind was designed for use only in a shallow prairie lake, never in water more than chest-deep; and that only the craft itself did the floating, never the photographer himself. I merely did the pushing and the piloting.

In chapter 7, "Saskatchewan," I recall some of my rather comical adventures in this lacustrine enterprise.

■

Here, finally, are a few notes on the 35mm equipment I used, and on the photographs of birds in general.

The particular camera systems I used, for what it may be worth, were Canons. With my special flash outfit I used only the old mechanical "F-1" cameras, with "FD" macro lenses (50mm f4, 100mm f4, and 200mm f4).

For a time I used this old F-1 system from the floating blind, as well, with a 400mm f4.5 lens; but recently I've used a more modern auto-focus "EOS" system, with a 500mm f4.5 lens. I haven't used digital cameras for any of these photographs.

The two slow Kodachromes, 25 and 64, are my eternal favorites, especially for use with the soft-flash system. They are sharp, smooth, natural in their color rendering—I dislike the candied colors, especially greens, of the other slow-speed films—and they are long-lived, as anyone knows who has gone through the old family slides and found those Kodachromes still bright and true after nearly half a century, even when they have been stored in less than kind conditions.

When photographing from the floating blind in natural light, however, I've used primarily the faster Ektachrome 100 films.

■

None of the photographs are of captive birds, if it need be said, or of birds restrained or baited in any way, except for the careful use of tape-recorded calls in one instance (the black rail). No props or perches were placed in the pictures, nor vegetation pulled up or cut away for a clear view; all scenes are just as they occurred in nature, except that at certain nests (American bittern, black rail) the foreground cover was held back temporarily, then brushed back into place.

Bibliography

Abbey, Edward, and Eliot Porter. 1973. *Appalachian Wilderness: The Great Smoky Mountains.* New York: Ballantine Books. Source of quotation on p. 82.

Armistead, Henry T. 1999. Maryland's Everglades: Southern Dorchester County. *Birding* (April): 141–154.

(Author Unstated). 2001. Oregon's Water War. *New York Times* (July 15), section 4: 14. Source of quotation on p. 108.

Bent, Arthur Cleveland. 1901. Journal. May 27–June 15. Unpublished. Arthur Cleveland Bent Ornithological Papers, 1880–1942. University of Massachusetts Amherst. W. E. B. Du Bois Library. Special Collections and Archives.

———. 1901, 1902. Nesting Habits of the *Anatidae* in North Dakota. *Auk* 18: 328–337; 19: 1–13, 164–174.

———. 1903. A North Dakota Slough. *Bird Lore* 5: 146–151. Source of quotation on p. 136.

———. 1905. Journal. May 24–June 17. Unpublished. Arthur Cleveland Bent Ornithological Papers, 1880–1942. University of Massachusetts Amherst. W. E. B. Du Bois Library. Special Collections and Archives.

———. 1906. Journal. June 1–July 1. Unpublished. Arthur Cleveland Bent Ornithological Papers, 1880–1942. University of Massachusetts Amherst. W. E. B. Du Bois Library. Special Collections and Archives.

———. 1907, 1908. Summer Birds of Southwestern Saskatchewan. *Auk* 24: 407–430; 25: 25–35.

————. 1919. *Life Histories of North American Diving Birds.* U.S. National Museum Bulletin 107.

————. 1921. *Life Histories of North American Gulls and Terns.* U.S. National Museum Bulletin 113. Source of quotations on pp. 142, 143–144.

————. 1923. *Life Histories of North American Wild Fowl* (part 1). U.S. National Museum Bulletin 126.

————. 1925. *Life Histories of North American Wild Fowl* (part 2). U.S. National Museum Bulletin 130.

————. 1927. *Life Histories of North American Marsh Birds.* U.S. National Museum Bulletin 135.

————. 1958. *Life Histories of North American Blackbirds, Orioles, Tanagers, and Allies.* U.S. National Museum Bulletin 211. Source of quotations on p. 137.

Bisport, Alan. 2000. Marshes Under Seige from Native Reeds. *New York Times* (October 8), Connecticut section: 10,

Brewer, T. M. 1875–1876, 1876–1878. Notes on Seventy-nine Species of Birds Observed in the Neighborhood of Camp Harney, Oregon, Compiled from the Correspondence of Capt. Charles Bendire, 1st Cavalry USA. *Proceedings of the Boston Society of Natural History* 18: 153–168; 19: 109–149. Source of quotation on p. 104.

Burt, William. 1994. *Shadowbirds.* New York: Lyons and Burford.

————. 1995. Stalkers of the Marsh. *Smithsonian* 26/2 (May): 98–103.

————. 2001. *Rare and Elusive Birds of North America.* New York: Rizzoli/Universe.

Chapman, Frank M. 1907, 1908. *Camps and Cruises of an Ornithologist.* New York: D. Appleton and Co.

Clark, John N. 1884. Nesting of the Little Black Rail in Connecticut. *Auk* 1: 393–394.

————. 1897. The Little Black Rail. *Nidologist* 4: 96–99.

Dahl, Thomas E. 1990. *Wetlands—Losses in the United States, 1780s to 1980s.* Washington, D.C.: U.S. Fish and Wildlife Service Report to Congress.

Dahl, Thomas E., and Gregory J. Alford. 1997. *History of Wetlands in the Coterminous United States.* Washington, D.C.: U.S. Geological Survey Water Supply Paper 2425.

Ducks Unlimited Canada. Undated. *Executive Summary, Crane Lake Project.*

————. Undated. *Completed Project Sheet, Crane Lake Project.* (File #67-30.)

Finley, William L. 1923. The Marshes of the Malheur. *Nature Magazine* 1/4: 46–48. Source of quotations on pp. 104, 105.

————. 1923. Hunting Birds with a Camera. *National Geographic* (August): 161–201. Source of quotations on pp. 103, 104.

Floor, Keith. 2003. Pots of Gold. *Audubon* 105 (December): 58–64.

Forbush, Edward Howe. 1914. The Sora Rail. *Bird Lore* 16: 303–306. Source of quotation on p. 2.

Fremont, John C. 1845. *Report of the Exploring Expedition to the Rocky Mountains in the Year 1842 and to Oregon and North California in the Year 1843–44.* Washington: Blair and Rives. Printed by Order of the House of Representatives. Source of quotations on pp. 120, 121.

Gates, David Allen. 1975. *Seasons of the Salt Marsh.* Greenwich, Connecticut: Chatham Press.

Godfrey, W. Earl. 1950. *Birds of the Cypress Hills and Flotten Lakes Region, Saskatchewan.* Bulletin No. 120, Biological Series. No. 40. Ottawa: National Museum of Canada.

Gorman, James. 2002. Aggressive and Invasive Reed Raises Concern. *New York Times* (March 17): www.nytimes.com.

Haramis, Michael, and Robert Colona. Undated. *The Effect of Nutria (Myocastor coypus) on Marsh Loss in the Eastern Shore of Maryland: An Enclosure Study.* Laurel, Maryland: USGS Patuxent Wildlife Research Center.

Hey, Donald L., and Nancy S. Philippi. 1999. *A Case for Wetland Restoration.* New York: John Wiley and Sons.

Hotchkiss, Neil. 1972. *Common Marsh, Underwater and Floating Leaved Plants of the United States and Canada.* New York: Dover Publications.

Houston, C. Stuart. 1982. Oology on the Northern Plains: An Historical Preview. *Blue Jay* 40: 154–157.

Jackson, Donald Dale. 1990. Orangetooth Is Here to Stay. *Audubon* 92 (July): 88–94.

Job, Herbert K. 1898. The Enchanted Isles. *Osprey* 3: 37–41.

———. 1902. *Among the Water-fowl.* New York: Doubleday, Page and Co.

Kerasote, Ted. 2001. Running on Empty. *Audubon* 103 (September): 23.

Line, Les, ed. 1990. The Last Wetlands. Special issue, *Audubon* 92 (July).

Luoma, Jon R. 1985. Twilight in Pothole Country. *Audubon* 87 (September): 66–84.

Mathewson, Worth. 1986. *William L. Finley: Pioneer Wildlife Photographer.* Corvallis: Oregon State University.

McKee, Russell. 1974. *The Last West: A History of the Great Plains of North America.* New York: Crowell.

Meanley, Brooke. 1975. *Birds and Marshes of the Chesapeake Bay Country.* Centreville, Maryland: Tidewater. Source of quotation on p. 79.

———. 1981. *Birdlife at Chincoteague.* Centreville, Maryland: Tidewater.

Mitchell, H. Hedley. 1924. Birds of Saskatchewan. *Canadian Field Naturalist* 38: 101–118.

Mitchell, John G. 1992. Our Disappearing Wetlands. *National Geographic* (October): 3–45.

Munroe, J. A. 1929. Glimpses of Little-known Western Lakes and Their Bird Life. *Canadian Field Naturalist* 43: 70–74.

Nadis, Steve. 1999. When It Comes to Building New Wetlands, Scientists Still Can't Fool Mother Nature. *National Wildlife* 37: 14–15.

Ness, Erik. 2001. Restoration: Murky Deal. *Audubon* 103 (November–December): 13. Source of quotation on p. 116.

Niering, William. 1966. *The Life of the Marsh.* New York: McGraw-Hill.

———. 1997. *Wetlands.* New York: Alfred A. Knopf.

Peabody, P. B. 1905. The Nesting of the Yellow Rail. *Warbler* 1: 49–51.

———. 1922. Haunts and Breeding Habits of the Yellow Rail. *Journal of the Museum of Comparative Oology* 2: 33–44.

Pearson, T. Gilbert, ed. [1917] 1936. *Birds of America.* New York: Garden City Books.

Peterson, Roger Tory. 1962. *How to Know the Birds.* Boston: Houghton Mifflin. Source of quotations on p. 134.

Petry, Loren C., and Marcia G. Norman. 1963. *A Beachcomber's Botany*. Greenwich, Connecticut: Chatham Press.

Potter, Lawrence E. 1930. Bird-life Changes in Twenty-five Years in Southwestern Saskatchewan. *Canadian Field Naturalist* 44: 147–149.

Riley, Laura, and William Riley. 1992. *Guide to the National Wildlife Refuges*. New York: Macmillan General Reference.

Ripley, S. Dillon. 1988. Revealing the Secret Lives of the Little Phantoms of the Marshes. *Smithsonian* 19/5 (September): 38–45.

Saltonstall, Kristin. 2002. Cryptic Invasion by a Non-native Genotype of the Common Reed, *Phragmites australis*, into North America. *Proceedings of the National Academy of Sciences* 99/4 (February 19): 2445–2449.

Smithsonian Institution Archives, Arthur Cleveland Bent Papers, c. 1910–1954. Washington, D.C.

Stansbury, Howard. 1852. *Exploration of the Valley of the Great Salt Lake of Utah Including a Reconnaissance of a New Route Through the Rocky Mountains*. Printed by Order of the Senate of the United States. Philadelphia: Lippincott, Grambo and Co. Source of quotation on p. 121.

Stein, Theo. 1994. *Phragmites:* Common Reed to Super Weed. Middletown, Connecticut: *Middletown Press* (August 1): A1, 8.

Steinhart, Peter. 1990. No Net Loss. *Audubon* 92 (July): 18–21.

Stevens, William K. 1995. Restoring Wetlands Could Ease Threat of Mississippi Floods. *New York Times* (August 8): C1, 4.

Taber, Wendell. 1955. In Memoriam: Arthur Cleveland Bent. *Auk* 72: 332–339.

Teal, John, and Mildred Teal. 1969. *Life and Death of the Salt Marsh*. New York: Ballantine Books.

Teale, Edwin Way. 1946. A. C. Bent: Plutarch of the Birds. *Audubon* 48: 14–20.

Tiner, Ralph W., Jr. 1984. *Wetlands of the United States: Current Status and Trends*. Washington, D.C.: U.S. Fish and Wildlife Service.

———. 1987. *A Field Guide to Coastal Wetland Plants of the Northeastern United States*. Amherst: University of Massachusetts Press.

U.S. Fish and Wildlife Service. Brochures (authors unstated):

1994. *Benton Lake National Wildlife Refuge.*

1994. *Grays Lake National Wildlife Refuge.*

1995. *Bear River Migratory Bird Refuge.*

1996. *Bear Lake National Wildlife Refuge.*

1999. *Klamath Basin National Wildlife Refuges.*

1999. *Malheur National Wildlife Refuges.*

Vileisis, Ann. 1997. *Discovering the Unknown Landscape: A History of America's Wetlands*. Washington, D.C.: Island Press.

Weske, John S. 1979. An Ecological Study of the Black Rail in Dorchester County, Maryland. Master's thesis, Cornell University.

Acknowledgments

I'm grateful to Jean Thomson Black, at Yale University Press, for her helpful spirit and enthusiasm for this project; and to Roland C. Clement, for introducing me to her. And I'm grateful to the other very helpful people at the Press, including Laura Davulis, Jessie Hunnicutt, and Kim Hastings.

Thanks to Brooke Meanley, for encouraging the idea of such a book in the first place, and for his insights on the marshes of the South, especially, which he knows well.

And thanks to Noble Proctor, naturalist extraordinaire, for his encouragement in this and other projects, and for his particularly keen and helpful commentary on the manuscript.

For their comments on the manuscript, or parts thereof, thanks to Henry T. ("Harry") Armistead, naturalist of the Maryland Eastern Shore; to Dave Mauser, at Klamath Basin National Wildlife Refuges in California; to Margaret Lowman, at New College of Florida; and to Laura Meyerson, at the University of Rhode Island.

For their observations on the ravaging invasive plant, *Phragmites,* I'm grateful to Paul Capotosto, of Connecticut's Department of Environmental Protection; to George O'Shea, at Prime Hook National Wildlife Refuge (Delaware); and to Don Temple, at Mattamuskeet National Wildlife Refuge (North Carolina). And I'm grateful to the following biologists, for their insights on other wetland issues: Gerry Deutscher, at Camas National Wildlife Refuge (Idaho); Richard Roy, at Malheur National Wildlife Refuge (Oregon); Dave Mauser, at Klamath Basin National Wildlife Refuges (California); Jon Hicks, at the Bureau of Reclamation in Klamath Falls (California); and Rob Bundy, at Bear Lake National Wildlife Refuge (Idaho).

Thanks to Judith Hassen at the Klamath County Museum, Klamath Falls, California, for providing records on water histories of the Klamath Basin Refuges.

For their help in the field, particularly in locating certain marsh birds, thanks to Doug Johnson, Ray Greenwood, and Don Petriezewski, all at Northern Prairie Research Station at Jamestown, North Dakota; Ron Bell and the late Mike Callow of Squaw Creek National Wildlife Refuge (Missouri); Carl and Linda Kurtz; Mike Male and Judy Fieth; Dave Lambeth; Judith Archer; Adam Burt, my son; and my dear late friend Carol H. Kimball, for her support and company on some recent camera travels.

Finally, I'm grateful to the late S. Dillon Ripley and the late Roger Tory Peterson, both, for encouraging my photography of birds.

■

Some of my photographs of the more elusive marsh birds in this book, and some of my descriptions of those birds in chapters 2 and 3, have appeared in various magazines over the years, including *Smithsonian* (August 1991, May 1995), *Audubon* (November 1982, September 1987), *National Wildlife* (February 2002), and *Living Bird* (Summer 1997). Some were featured also in my earlier books about elusive birds: *Shadowbirds* (1994), and *Rare and Elusive Birds of North America* (2001).

An adaptation of chapter 1, about Connecticut River marshes, appeared in *Connecticut Magazine* (September 2005).

Index

16, 117–22; in Canada, 137–48; in Louisiana, 81–83; in North Dakota, 133–37

Fuertes, Louis Agassiz, 101

fulvous whistling duck, 83

gadwall, 105, 139

gallinule, purple, 81, pl. 63

geese, 105, 106, 114

Georgia, and Sea Islands, 83, pl. 55

gerardias, 19

Godfrey, Earl, 140

godwits, 138, 139

goldenrod, 19, 44

Goose Island (Conn.), 17–18, 25

grackle, boat-tailed, 81

Gray's Lake National Wildlife Refuge (Idaho), 119

Great Bay (N.J.), 85–86, pl. 58, pls. 60–61

great blue heron, 105

great egret, 105

Great Island (Conn.), 19–21

Great Salt Lake (Utah), 120, 121

grebes, 104, 105, 135, 138, 147. *See also* eared grebe; pied-billed grebe; western grebe

gulls, 138, 140. *See also* California gull; Franklin's gull; laughing gull; ring-billed gull

Hackensack Meadows (N.J.), 85

herons, 81, 103, 105

high-tide bush, pl. 71

ibis, glossy, 81

ibis, white-faced, 119–20

irrigation farming: and marsh drainage, 106; and wildlife refuges, 117–19, 120

Job, Herbert K., 143

Johnston, Carol, 116

Juncus gerardi (black grass), 19

Juncus roemerianus (needlerush), 42

king rail, 19, 23, 83, pl. 7, pl. 9, pl. 21; nest and eggs; pl. 22

Klamath Basin (Calif. and Ore.), 103–4, 106–7, 108, 110, 111

Klamath lakes (Calif. and Ore.), 103–4, 106–7, 109, 110–11

Klamath marshes, 84, 106–7, 111

Klamath Reclamation Project (Calif. and Ore.), 106, 108–9, 119

Klamath River, 103

knots, 84

land development, 114–15

large blue flag, 82, pl. 5

laughing gull, 81, 83–84, pl. 64; nest and eggs, pl. 29

least bittern, 18, 22, pl. 8, pls. 10–12, pls. 24–26

LeConte's sparrow, 62, 63, pl. 47

lily, yellow pond, pl. 76

lily pads, pl. 53

Locassine National Wildlife Refuge (La.), 83

loosestrife, purple (*Lythrum salicaria*), 3, 22

Louisiana coastal marshes, 80–83

Lower Suwanee National Wildlife Refuge (Fla.), 83

Malheur Lake (Ore.), 104–6, 107–8

Malheur National Wildlife Refuge (Ore.), 107–8

mallard, 105, 114, 135, 138–39

market hunting, 104

marsh fern, pl. 38

marsh hen, 83

marsh marigold, pl. 17

marsh wren, 18, 22, 43, 44, 135

Maryland, and Elliott Island marsh, 41–46, pls. 27–28, pl. 32

Mattamuskeet National Wildlife Refuge (N.C.), 23

meadowlark, 42, 44

Meanley, Brooke, 79

milkweed, coast, pl. 3

Scott Provost

William Burt is a naturalist, photographer, and writer with a passion for wild places and elusive birds. He is the author of two previous books, and his photographs and stories have appeared in *Smithsonian, Audubon, National Wildlife,* and other magazines.